1·75

AIRFIX
magazine guide 8

German Tanks
of
World War 2

Terry Gander and
Peter Chamberlain

Patrick Stephens Ltd
in association with Airfix Products Ltd

First published — 1975
Reprinted — 1978

ISBN 0 85059 211 9

Don't forget these other exciting titles!

No 1 *Plastic Modelling,*
by Gerald Scarborough
No 2 *Aircraft Modelling,*
by Bryan Philpott
No 3 *Military Modelling,*
by Gerald Scarborough
No 4 *Napoleonic Wargaming,*
by Bruce Quarrie
No 5 *Tank & AFV Modelling,*
by Gerald Scarborough
No 6 *RAF Fighters of
World War 2,*
by Alan W. Hall
No 7 *Warship Modelling,*
by Peter Hodges

Cover design by Ian Heath

Text set in 8 on 9 pt Helvetica Medium
by Blackfriars Press Limited,
Leicester.
Printed on 85 gsm Factotum Cartridge
and bound by the Garden City Press,
Letchworth, Herts.
Published by Patrick Stephens Ltd,
Bar Hill, Cambridge, CB3 8EL, in
association with Airfix Products Ltd,
London SW18.

Contents

Acknowledgements

We would like to thank the Tamiya Mokei company of Japan and Richard Kohnstam Ltd for permission to use the drawings on pages 21, 25, 36, 41 and 45, and Gerald Scarborough for supplying those on pages 18 and 33.

Editor's introduction

Of all the fighting weapons which have existed throughout mankind's long and bloodthirsty history, none exercises more fascination than the tank. In particular, the tanks used by the German army during the last World War have acquired a patina of glamour with a singular appeal to modern imagination.

Volumes have been, and continue to be, written on the subject of German tanks and their role in the last war, ranging from large, lavishly illustrated and expensive volumes to small paperback booklets on individual tank models. Never before, however, has all the basic information on German tank development, armament, nomenclature, camouflage and markings been collected together into one concise but detailed reference guide at a price within anyone's pocket.

Newcomers to the subject of German 'panzers' can learn from this book the origins of German armoured theory; how and why the tanks were designed as they were; what their capabilities, performance, armour plate and guns were like; how they were used and how they were painted.

'Old hands' will also find the book an invaluable ready-reference to all the German-designed and built tanks used during World War 2. The performance data will be extremely useful to wargamers seeking to compile realistic playing rules for World War 2 combat, while the 1:76 scale drawings and chapter on camouflage and markings will enable any modeller to complete accurate models of the vehicles in this popular scale.

The authors really need no introduction, since both are established authorities on German wartime weaponry and have contributed frequently to *Airfix Magazine* over the years, and Peter Chamberlain has specially selected the photographs for this book — most of which have never been published before.

Practically every variant of every tank, from the diminutive PzKpfw I to the mammoth 'King Tiger', is described and illustrated, including special-purpose versions and self-propelled guns. At the end is a basic book list for readers who wish to delve more deeply into this complex and fascinating subject.

Future *Airfix Magazine Guides* will cover the tanks of other combatant nations in the same way.

BRUCE QUARRIE

one

German tank developments to 1939

For a nation that was usually in the technical forefront of the art of war, Germany was strangely late in the development of the tank. It took the arrival of the first British and French tanks on the Western Front battlefields to give an impetus to the search for a form of mobile armoured fire platform among the German armament designers, but by the time the need for a German answer to the Allied tanks was shown, German industry was already at full stretch and very few German tanks were produced prior to the end of 1918.

The only German tank actually to see service was the awkward A7V. This was a large armoured box built onto a modified Holt tractor chassis. Even when compared with the slow, lumbering and unreliable British tanks, the A7V was at a disadvantage. It was high, awkward, slow and vulnerable and required a crew of no fewer than 18 men. Very few were produced and the bulk of German tank units were made up from captured British tanks (beutepanzer). Improved and better German designs did reach the project stage but none saw service before the German defeat of November 1918.

Under the terms of the Versailles Treaty of 1919, the German Army was limited to 100,000 men and they were not allowed a tank arm. All German tanks then existing were either scrapped or carted off to museums, leaving the German Army with only a few armoured cars and tenders. But in many ways the small size of the new German Army was an advantage. With so few men under arms, the accent was on quality, both in training and methods. By keeping the length of time that volunteers could serve down to a minimum, a sizeable trained reserve of men was soon formed, and the tactics that this army was to use were closely studied by some of the best officers that Germany could produce.

One of the most open minds of the German Army during the 1920s belonged to the man who was later to become chief of the German 'panzer' (tank) arm, namely Guderian. He closely studied the lessons of World War 1, and read widely to learn of possible methods of overcoming the static conditions of the Western Front.

Dummy tanks built on to light motor cars being used on exercises during the late 1920s.

Above *The Panzerkampfwagen I Ausf A — the first of the German tanks produced after Hitler came to power.* **Below** *A PzKpfw 38(t) in action in France, 1940.*

His studies began in 1922, and at about the same time the need for a small powerful army was being proposed by the Chief of the German General Staff, von Seeckt, another brilliant officer who laid the foundations for the future Wehrmacht (German army). These two officers were in the vanguard of a faction that changed the future of warfare for between them they carried out the proposals of the British tactical prophets, Fuller and Liddell Hart.

These two military thinkers produced a series of writings on warfare which went against all that had gone before in that they proposed that the war of the future would be fought by highly-mechanised forces based on the power of the tank. Concentration of striking forces (the 'schwerpunkt') at a weak point would produce a breakthrough which would be exploited by the rapid concentration of all forces into what was known as the 'expanding torrent' which would penetrate deep into the enemy rear, and disrupt communications and supply routes. Such revolutionary tactics took little root in Britain or elsewhere, but they were just what was needed in Germany and the German Army Staff officers began the slow task of preparing for a mobile war based on armoured units. A gradual programme of training and of close co-operation between the various service arms began in about 1925.

The main snag to the ambitious proposals was that the German Reichswehr had no armoured vehicles other than a few armoured cars to experiment with. The first full-scale manoeuvres involving mechanised forces took place in 1926, but in place of tanks men were employed to carry cardboard silhouettes of tanks and some motor cars were used with card or timber hulls resembling tanks. Perhaps this was the source of the 'German cardboard tank' rumours which were prevalent in Britain in 1939. But this lack of vehicles did not indicate that no research was being carried out during the 1920s into tank production. The truth was that Germany had been carrying on clandestine research from about 1920 onwards, despite the strictures of the Versailles Treaty forbidding such activities.

During the early 1920s, German designers had been active in Sweden and had gained experience in the production of a light tank based on the LK II design of 1918. A small batch of these vehicles was produced for the Swedish Army but none for Germany. Back in Germany, the General Staff issued a secret specification to German industry to produce prototypes of two types of tank. One was intended as a light tank of about nine tons mounting a 3.7 cm gun in a turret. The other type was seen as a medium tank with a 7.5 cm gun and weighing about 20 tons. This latter vehicle was very well armed for its time and in design concept was very advanced.

Both prototypes were built and tested on the Russian facilities at Kazan in Russia as a result of a political agreement, and the trials were conducted under great secrecy in 1928. As a result of the trials of the 'Grosstraktor and 'leichtetraktor', as the two designs were code-named for security purposes, a further two designs were proposed to be known as the Neubaufahrzeuge, but these

could not be built until the mid-1930s. This delay factor in producing tanks for the German Army was one of the main lessons learned during the early experiments. It became apparent that the production of modern tanks was going to involve a great deal of industrial and development potential before the needs of the German Army were to be met.

By 1930 the need for some form of tank for tactical trials and training was becoming urgent. A possible answer seemed to be the light machine-gun carrier based on the design of the British Carden-Loyd. This type of vehicle could be produced relatively easily and quickly, and by 1932 the industrial potential had been developed to the standards needed to build such a vehicle in quantity.

In 1933 the Nazi Party came to power and all pretensions of adhering to the terms of the Versailles Treaty were set aside. In that year orders were placed for a light tank weighing 5.3 tons and mounting two machine-guns in a small turret. The crew was to be two men.

This vehicle emerged as the Panzerkampfwagen I (PzKpfw I), built by Krupp. It was built in two main versions and despite the intention to use it as a training vehicle only, it saw action both in the Spanish Civil War and during the early stages of the Second World War. Its main task was as a trials vehicle for tactical experiments and it was also used as a propaganda vehicle in a long series of parades and mock battles that did much to bolster the illusion of German strength both at home and abroad.

At the same time as the specification for the Panzer I was issued, an order for a slightly larger vehicle was also given. It was intended to be a three-man tank armed with a 20 mm cannon and a machine-gun, and it was to be a more battle-worthy vehicle than the little PzKpfw I. A series of prototypes was produced, but in 1934 the MAN (Maschinenfabrik Augsburg Nürnberg) version was chosen for production as the PzKpfw II. The PzKpfw II in its original form resembled a scaled-up PzKpfw I, but it was heavier and later versions used a revised suspension. In service it was used as a reconnaisance vehicle, but its main disadvantage was its small gun which was to prove too light for armoured combat.

The PzKpfw I and II were produced in large numbers, and were the mainstays of the German panzer arm up till 1939 and during the early war years. It was with these vehicles that the panzer divisions trained and prepared themselves for war.

German war plans envisaged that war was unlikely before 1942, by which time heavier tanks would be in service. As we have seen these tanks were proposed as far back in time as 1925 but it was not until 1935 that the first orders for a heavy tank were issued.

This heavy tank was intended to have a heavy 7.5 cm gun, weigh about 20 tons and have a crew of five men. Its true intention was disguised by the

Above *Panzerjäger 38(t) mit 7.5 cm Pak 40/3 Marder III — one of many types of self-propelled gun mountings based on the PzKpfw 38(t) chassis.* **Below** *Jagdpanzer 38(t) Hetzer. This vehicle was one of the best of the many types of tank-hunting vehicles used by the Germans during World War 2. It was once again based on the PzKpfw 38(t) chassis.*

German tanks of World War 2

designation of 'battalionsfuhrer-wagen' (battalion commander's vehicle) and the usual series of prototypes was produced before the Krupp vehicle was awarded the production contract. The result was the PzKpfw IV, and it became the most widely used and encountered of all the German tanks. In time, it ran to a long series of sub-marks (described in Chapter 4) and was progressively up-gunned and up-armoured as the tactical situation demanded.

At first, production was slow and development was rather protracted due to the industrial effort involved and the general lack of experience in producing tanks, but the PzKpfw IV became one of the most dependable vehicles in use with any army involved in World War 2. On September 1 1939 there were only 211 PzKpfw IVs available when Germany invaded Poland.

In 1936 it was decided to order another battle tank, slightly smaller than the PzKpfw IV and known as the 'zugfuhrerwagen' (platoon commander's vehicle). Armed with a 3.7 cm gun, this tank was meant to supplement the PzKpfw IV, and the end result, the PzKpfw III, resembled a small PzKpfw IV. In time, the PzKpfw III was produced in large numbers and also ran to a long series of sub-marks. It too, was progressively up-gunned and fitted with thicker armour as the war progressed. Early production was slow, so slow that on September 1 1939, only 98 PzKpfw IIIs were available for use.

The slow production of the PzKpfw III and IV meant that large numbers of the light PzKpfw Is and IIs were still in front-line service at a time when they should have been phased out in favour of new tanks.

In 1939 this situation was considerably alleviated by the annexation of Czechoslovakia, and the tank park of the powerful Czech Army was transferred to the German Army. Just as important as the 469 battle-worthy tanks that were taken over was the important Skoda Works at Pilsen which continued to produce tanks and armaments for the German war effort.

The Czech tanks taken over by the Wehrmacht were of two types, the LT-35 and the LT-38. They were impressed into the panzer divisions as the PzKpfw 35(t) and PzKpfw 38(t) respectively — (t) stands for tscheschisch, namely Czech. Both mounted 37 mm guns, but the most important type was the PzKpfw 38(t) which was to become one of the most successful tank chassis ever designed, for as well as using the type as a battle tank, the Germans used the chassis as the basis for a long string of self-propelled artillery, tank hunter and assault gun vehicles.

When war came in 1939 it came some three years too early for the German General Staff plans. Their ideas on warfare were based on well-armoured, well-armed tanks, but in 1939 only a few of the German tanks in service could be so described. But the German Army was superbly trained, in excellent fighting condition and ready for anything. The 'Blitzkrieg' warfare was unleashed on Poland on September 1 1939.

two

World War 2

While the world watched, Germany defeated Poland in 18 days. This astounding success, even against a relatively unprepared nation like Poland, was an outstanding example of the new mobile warfare predicted by Fuller and Liddell Hart during the 1920s. Tank divisions massed at a few points, concentrated all their forces at one objective, broke through and immediately struck deep into the enemy rear.

The main striking force came from the massed tanks of the panzer divisions, supported by artillery at first, but as the tanks penetrated deeper into Polish territory, support came from the Luftwaffe, and especially from the Junkers Ju 87, the infamous 'Stuka'. Motorised infantry was bought up to actually occupy the territory taken and to reduce any strong points left behind by the 'expanding torrent'.

All arms of the German Wehrmacht worked together with a cohesion impossible in many armies, but the main striking forces were concentrated in the panzer divisions. In these novel formations tanks were supported by their own artillery, engineer, supply and signal formations, together with their own infantry units, later to become the famous 'Panzergrenadiers'. The panzer division was the living embodiment of many of the early military prophets but it took the peculiar history and political situation of Germany to put the idea into life.

After the conquest of Poland the panzer divisions took stock and re-organised themselves for the next major campaign, which was to be in the west. For this campaign, new equipment was issued in the shape of increased numbers of PzKpfw IIIs and IVs, and more Czech equipment became available. The campaign in the west was approached with more caution than had been necessary in the Polish conflict as France was, on paper at least, one of the most powerful nations in Europe.

The number of tanks in the French Army was almost treble in number to those of the Wehrmacht, and their quality was also on a parity. But the events of May 1940 proved once again the value of the German tactics and training. By the end of June France had been conquered and the little British contingent had been forced to flee from Dunkirk back to Britain.

Once again, deep armoured thrusts had swept through the French rear and destroyed communications, supply routes and, what was to be the deciding factor, they destroyed the French morale and will to fight. A few determined counters were delivered by Allied tank formations but the panzer divisions otherwise had it all their own way, and France surrendered. This was the high-water mark of the panzer divisions and the 'Blitzkrieg'.

Throughout the early campaigns of 1939 and 1940, the equipment of the panzer divisions remained unchanged from the original plans. The PzKpfw I, which had been pressed into service, showed itself incapable of standing up to the demands made upon it but as it was designed as a training vehicle this is not surprising. After 1940 it was gradually withdrawn from use as a front line tank and was used for driver training, as a munitions carrier, command vehicle or tractor, and for various special engineer vehicles.

The PzKpfw II had shown itself to be a useful reconnaissance vehicle but its armament of one 20 mm cannon was too light for anything else. Heavier armour was fitted to later variants but the type was gradually produced in smaller numbers and replaced by heavier tanks. One later variant that

A PzKpfw III Ausf E.

was produced in some numbers was the PzKpfw II Ausf L, known as the Luchs (Lynx). Production of this variant began in late 1942, and the design featured heavier armour, revised suspension and some were fitted with a 5 cm gun, although most retained the 20 mm cannon. A total of 647 PzKpfw IIs of all types was produced.

In May 1940 there were 329 PzKpfw IIIs ready for use, and in France they proved to be a most useful tank, but it was felt that more armour and a more powerful gun were needed. It was at this point that Hitler took a personal interest in the panzer arm. He ordered that a 5 cm L/60 gun should be installed in future variants of the PzKpfw III, but for various reasons this was altered to an L/42 gun. In explanation, the 'L/' denotes the length of the gun expressed in calibres, eg L/42 means that the gun is 42 times the calibre long. Thus the L/42 was 50 × 42 mm, or 2,100 mm long.

The longer a gun barrel is, the higher the muzzle velocity and thus the striking power, so an L/60 gun would be more powerful as an anti-tank gun than an L/42 weapon.

As things were to turn out, the L/60 was not fitted before the Russian campaign started, and the lack of it was to have severe effects on the use-fulness of the PzKpfw III. When Hitler discovered that his order had not been carried out, he was furious and from then on he personally supervised the armament and development of German tanks to the extent that his 'intuition' often overrode more practical changes, and led to some unfortunate decisions.

In time the PzKpfw III was fitted with the L/60 gun and was eventually fitted with the low velocity 7.5 cm gun fitted to the original six PzKpfw IV versions. Production of the PzKpfw III ceased in 1943 but by that time a considerable number of PzKpfw III hulls were being diverted towards the 'Sturmgeschutz' assembly lines.

These assault gun carriages first took shape during the 1940 France campaign when a number of PzKpfw I chassis were used to carry 15 cm sIG 33 guns. They had the advantage over conventional tanks of being cheap and easy to produce and, after 1940, captured tank chassis that could not be used as panzer division equipment could often be diverted for the mounting of anti-tank or artillery pieces to bulk out panzer units. The main disadvantage of this philosophy was that such assault guns lacked the vital 360° turret traverse essential in armoured warfare.

Gradually, increasing numbers of PzKpfw II and III chassis were diverted from tank production towards the Sturmgeschutz lines, and the tank content of panzer divisions suffered as a result. In addition, some numbers of PzKpfw III tanks were diverted towards such tasks as command tanks, mobile observation posts for artillery, and flamethrowers (flammpanzer).

The PzKpfw IV had proved itself a most battle-worthy tank in France, and went on to further establish itself in Greece and the Western Desert. It formed the backbone of the panzer divisions throughout the war, but after 1940 it was progressively up-armoured and the gun was replaced by a more powerful L/43 weapon and eventually by the very successful L/48 version which could outrange and outfight nearly all its contemporaries.

Meanwhile, the encounters that the panzer units fought with such vehicles as the British Matilda and the French Char B during the 1940 campaign had shown that Germany had tended to sacrifice striking power for armour, with the exception of the PzKpfw IV. Hitler himself took a hand in future equipment trends and insisted on a new heavy tank with sufficient armour and armament to take on any possible tank it was likely to encounter.

Orders for prototypes were put out in May 1941, and there were two main contenders. One was a complex design by Porsche which was rejected only after 90 had been built. The hulls were converted to assault guns and became the 'Elefant', which was one of the German armament industry's greatest failures. The other contender was a Henschel design which became the PzKpfw IV 'Tiger'.

When it first appeared in March 1942 it tipped the scales at 55 tons and was thus the world's heaviest tank in service. It had a thick armoured hide and what was then the remarkably heavy armamemt of the 8.8 cm KampfwagenKanone (KwK) 36, a development of the 8.8 cm Flak 18 anti-aircraft gun. In addition, two machine-guns were fitted.

The Tiger was a considerable prob-lem for Allied commanders to counter, and when it was first used in Tunisia in 1943 it was only defeated with a great deal of difficulty. But for all its fighting merits the Tiger was not a very suc-cessful fighting tank. Its weight and bulk made it a very slow and awkward vehicle to employ. Its armamemt could pick off potential enemies at very long ranges but in close fighting its slow rate of turret traverse placed it at a considerable disadvantage.

Perhaps its main disadvantage was its lack of mechanical reliability. It had been pressed into service when many of its mechanical components had not been fully developed, and the result was a very low mechanical reliability factor. It was also very expensive, cost-ing 250,800 RM (Reichs Marks) as opposed to 103,462 RM for a PzKpfw IV. It went out of production during 1944 but up till then the Tiger was always used as a spearhead of the panzer divisions, and was usually issued to elite formations only.

By the time the Tiger had come into service, the Russian campaign of 1941 had begun. Full of confidence and with victorious campaign experience behind them, the panzer divisions swept across the Russian steppes, duplicating over and over again their heady victories of 1939 and 1940.

When the campaign began in June 1941, the Germans had at their dis-posal 5,264 tanks of all types of which about 3,350 were in the front line (a few of these continued to be the little PzKpfw I), of which the bulk were PzKpfw IIIs and IVs.

Five months after that the panzer divisions were deep in Russia and had captured or destroyed over 17,000 Russian tanks, which was almost the entire Russian tank strength. But the Germans had also come up against what was to prove one of the most remarkable weapons of World War 2, namely the Russian T-34/76 tank. As soon as it was encountered the Ger-mans realised that their own tanks would be inadequate against large numbers of this Russian product.

The T-34 had well-sloped armour which tended to deflect solid shot, a

German tanks of World War 2

powerful 76.2 mm gun, a good turn of speed, and was potentially available in huge numbers. The only German vehicles that could encounter it were the PzKpfw IVs armed with the L/48 7.5 cm gun and there were not many of them in service in late 1941. The PzKpfw III should have been able to counter the armour of the T-34 if it had been fitted with the 5 cm L/60 as ordered by Hitler, but none of these would be ready until mid-1942.

An emergency specification based on the T-34 was rushed out to German industry. Many firms favoured a direct copy of the T-34, but in the nationalistic Nazi state of 1941 this was politically unthinkable. The accepted design was produced by MAN who designed a vehicle that was to gain fame as the PzKpfw V 'Panther'.

The Panther was the most successful all-round battle tank to be designed in Germany. It was armed with a potent 7.5 cm L/70 gun and it featured well-sloped armour and torsion bar suspension. For its size it was rather heavy at 43-45 tons, but it had a good turn of speed and was manoeuvrable and handy. It was not ready for action until 1943 but until then the PzKpfw III and IV had to counter the T-34 and its heavier partner, the KV-I, alone.

Once again tactics and fighting skill led the German panzers on during 1942 but by the end of the year, the Stalingrad defeat marked the end of the Wehrmacht advances. From that time on, the initiative passed to Russia and her allies and apart from local successes the panzer divisions were on the defensive. The 'Blitzkrieg' era of rapid and total victories had passed, and the war turned into a bloody slugging match on all fronts.

The passing of the era of the tank's supremacy was marked by the Battle of Kursk in 1943. Kursk was the greatest land battle of all time, and was fought by tank armies, instead of the usual divisions. It was a battle launched against a large Russian salient in central Russia by the German tank armies during July 1943. The Germans placed great reliance on the new Panther tank and its heavy counterpart, the Tiger. The start of the battle had been delayed by the Germans in order to get enough Panthers into the line, but the result was a disaster for the Germans.

Their attack was launched against carefully and heavily defended localities, and this time there was no

Captured examples of French tanks being used for driver and crew training in France. The vehicle in front is a Somua S35 (PzKpfw 35-S 739(f)) followed by a Hotchkiss H39 (PzKpfw 39-H735(f)).

armoured break-through. The panzers were halted by a ferocious defence in depth, and in addition large numbers of the under-developed Panthers and Tigers which had been rushed into battle simply broke down and were lost to tank-killer squads. It was a heavy defeat for the Wehrmacht and thereafter they began to fall back towards Germany.

The panzer divisions fought as hard as ever but they were nearly always on the defensive. The bulk of their formations continued to use the faithful PzKpfw III and IV, but increasing numbers of Sturmgeschutz were employed to plug the gaps made by increasing tank losses.

As time went on the tanks themselves took on a more defensive appearance. The arrival of the hollow-charge anti-tank device on the tactical scene meant that tanks had to carry stand-off armour in the shape of thin metal sheets held suspended from the sides of vehicles. The Germans called these sheets 'Scheutzen' (skirts), and they countered hollow-charge missiles by making the hollow

charge expend its energy by exploding away from the side of the tank itself. To counter anti-tank mines and charges placed on the tank itself by tank-killer infantry squads, the surfaces of German tanks were coated with 'Zimmerit', a plaster-like substance which prevented magnetic fixing devices from operating.

There was only one more major German tank to see service before the end of the war after the Panther and that was the mighty PzKpfw VI Tiger II, or Königstiger. This monster emerged from a specification intended for a Tiger replacement, and the first was ready by the end of 1943. It was not until the end of 1944 that the first Tiger II was issued to the panzer divisions.

The Tiger II weighed nearly 70 tons and was armed with a developed version of the 8.8 cm gun, namely the 8.8 cm KwK 43. It was a most remarkable piece of engineering produced under extreme difficulties brought about by constant air attack from Allied bombers, and nearly 500 were built.

It was a formidable fighting machine but again, its weight and bulk dictated

A PzKpfw IV Ausf F1 in Russia sporting a rough winter white camouflage.

German tanks of World War 2

that it was suitable for defensive fighting only. Also, it was mechanically under-developed and produced a rich crop of mechanical failures. Nevertheless, the appearance of a Tiger II on a battlefield put fear into many an Allied heart for it was a truly formidable opponent. Only the Russian Joseph Stalin I and II could have been anything like a match for it.

As the war ended, the old faithful, the PzKpfw IV, was still in production and action. The Panther had gradually taken over from the PzKpfw IV but had never replaced it, and the PzKpfw III had gradually been relegated to the role of infantry support tank. More and more Sturmgeschutz vehicles had taken over from tanks in the ranks of the attenuated panzer divisions, which by 1945 had become only a shadow of their former selves and were fighting not as divisions but in defensive battle groups formed to meet local conditions.

Mention must be made of the large numbers of captured vehicles used by the Germans. Any tanks that were captured were eventually used by the resourceful Germans in some role or other, usually in the mundane role of artillery tractor or as the carrier for some form of gun. Some tanks were used as front-line equipment.

The important Czech PzKpfw 35(t) and 38(t) have already been mentioned, but large numbers of French tanks were used by second-line units in France and Russia for occupying and police duties. Perhaps one of the most famous tanks taken into German service was the T-34. Large numbers of captured T-34s were turned against their former owners on the Eastern Front during 1942 and 1943 under the designation PzKpfw T34-747(r). In the Western Desert some numbers of British Matilda tanks became the Infantrie PzKpfw Mk II 748(e), and in North-West Europe many Shermans became the PzKpfw M4-748(a).

By 1944, drastic changes had been made to the methods of production. Despite Allied air attacks, more and more tanks were driven off the assembly lines, but instead of concentrating all possible resources on a few models, as was the successful Russian method, a growing number of different types were projected. A whole new family of different models was proposed at one point. This was the 'E' series which would have ranged from the E.5, weighing only five tons, up through a range of another four models to the monster 140-ton E.100. Of this range, only the massive E.100 got anywhere near the hardware stage and it was not completed before the war in Europe ended.

Other projects that did little to increase the number of tanks in the field were the odd proposal to build a 1,500 ton self-propelled 80 cm gun for street-fighting, and a series of huge mortars on self-propelled platforms. These weird and tactically almost useless schemes did much to divert design and production facilities away from such essential requirements as the need for more Panthers in the field.

Perhaps the most bizarre of all these diversionary projects was the unlikely 180-ton mammoth known as Maus (Mouse). This project was personally approved by Hitler and went ahead with no formal backing other than the Fuhrer's approval. The Maus mounted a 15 cm and 7.5 cm gun in a huge turret, and its weight and size meant that it was more of a mobile pill-box than a useful tank, but the project went ahead absorbing much design and manufacturing potential that could have been employed on more useful purposes. In the end, the Maus never saw action for the war ended when it was still under development.

The war ended with the once mighty panzer arm in disarray. Harried by constant air attack and virtually immobilised by lack of fuel, they were a mere shadow of their former selves. At the end, the PzKpfw IV was still in the line, and along with the Panther and the Tiger and Tiger II, held off the advancing Allies as long as possible — but the days of deep armoured thrusts and headlong pursuits into the enemy rear were over.

three

German tank designations

A brief word is needed to explain what would seem at first sight to be a rather complex method of designating vehicles used by the German Army.

To begin with, all vehicles were prefixed by their purpose. For tanks this was the word 'Panzerkampfwagen' which translates literally as 'armoured fighting vehicle'. This was usually abbreviated in practise to 'PzKpfw'. Other examples of this system were Panzerbeobachtungswagen (PzBeobWg), which was an armoured observation vehicle; Flakpanzer (FlakPz), which was an anti-aircraft tank; and Panzerbefehlwagen (PzBefWg), which was an armoured command vehicle.

In addition to the purpose prefix each type of vehicle was given an ordnance inventory number. This was the number used for labelling all working drawings and spare parts lists peculiar to that particular type of vehicle. This ordnance number was known as the 'sonderkraftfahrzeug' number, or SdKfz, and was usually included in brackets at the end of a vehicle designation. (SdKfz literally translated indicates 'special purpose motor vehicle'). An example would be PzKpfw I Ausf B (SdKfz 101).

The word 'Ausf' means Mark, being short for 'Ausführung', and could be found written either as Ausf or ausf. The different marks were written as letters of the alphabet. Thus for example, different marks of the Panther were described as the Ausf A, D or G.

The SdKfz numbers for each vehicle type are given in the data tables. Captured vehicles were not normally given an SdKfz number, exceptions being made for Czech vehicles that remained in production for the Wehrmacht, and for certain conversions such as those utilising the Lorraine tractor chassis as a self-propelled gun mounting.

(NB In this book the international system of classifying the calibre of guns in mm has been used for all foreign-made weapons. The Wehrmacht designated their weapons in mm up to 20 mm and in cm thereafter, eg 8.8 cm, and this system has been followed in classifying all German weapons. — Ed).

four **Models**

German tank data

PzKpfw I (SdKfz 101)

Origins

The PzKfw I had its origins in the design of the Carden-Loyd machine-gun carrier, an example of which was obtained for study during 1932. Five German firms submitted designs and the Krupp entry was awarded the production contract. The Krupp entry was designated the LKA I, but was given the cover name of La S, which stood for agricultural tractor. Krupp was responsible for the hull and chassis, and Daimler-Benz built the superstructure, but the first prototypes were built by Henschel and delivered during December 1933.

PzKfw I Ausf A The first production variant was the Ausf A and 150 were built by Henschel, starting in July 1934. Weight of the original PzKpfw I was 5.4 tons. It differed from the LKA I prototype in having smaller road wheels and an external beam supporting the suspension. Armour was 13 mm thick, and armament was two 7.92 mm machine-guns mounted side-by-side in the turret. The first Ausf A vehicles were delivered during 1935.

PzKpfw I Ausf B The Ausf B formed the major type of the 1,500 PzKpfw Is built. It was longer than the Ausf A due to the addition of an extra road wheel to improve traction, and the engine was improved from a 60 hp model to a 100 hp petrol engine. The armament and armour remained the same but weight was increased to 5.8 tons. This was the variant which formed part of the panzer divisions during the invasions of Poland and France, and a few were left in service in 1941 when Russia was invaded.

In 1939, a proposal was made to develop the PzKpfw I as a small infantry support tank. A few prototypes were built which featured thicker armour and a redesigned suspension. On one design the twin machine-guns

PzKpfw I Ausf B.

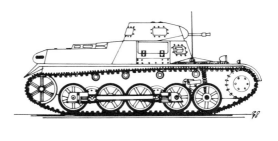

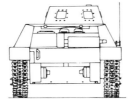

PzKpfw I
1:76 scale

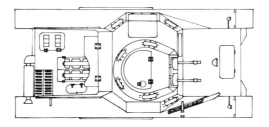

Kleiner Panzerbefehlswagen I captured by the British in North Africa.

German tanks of World War 2

were replaced by a 20 mm cannon with a co-axial machine-gun, but the project did not get beyond the prototype stage.

Variants

After 1940 the basic PzKpfw I was used as the basis for a series of special vehicles. The main variants were as follows:

Kleiner Panzerbefehlswagen I (SdKfz 265) On this variant the turret was replaced by a fixed box structure and the result was used as a mobile command post. One machine-gun was fitted.

Flammpanzer I These were Ausf A field conversions made in North Africa to give the Afrika Korps a mobile flamethrower. Only a few were so converted.

PzKpfw I(A) Munitions-Schlepper (SdKfz 111) This was a conversion of the Ausf A in which the turret and superstructure were removed to enable the open hull to be used for carrying ammunition for mobile columns.

15 cm sIG 33 auf Geschutzwagen I

Below 15 cm sIG 33 auf Geschutzwagen I Ausf B. **Bottom** Panzerjäger I für 4.7 cm Pak(t).

Ausf B Produced in 1939, this carried a modified infantry gun in a high open box. It saw action in 1940 and 1941 but was withdrawn soon after as the gun was really too heavy for the chassis. This was the first German self-propelled gun.

Panzerjäger I für 4.7 cm Pak(t) In order to give a degree of mobility to anti-tank units a number of Czech guns were placed on Ausf B chassis behind a shield installed in place of the turret. This variant became the first of the Panzerjäger tank-hunting vehicles which were later produced in a wide variety of types and calibres. The PzJäg I saw action in France and North Africa.

PzKpfw Ib Ladungswerfer 1 This was a specialised engineer vehicle used to carry demolition charges on a movable gantry. Not many were so converted.

Gutted hulls were often used as tractors or for driver training.

Data PzKpfw I Ausf B

Weight in action 6 tons

A PzKpfw 1 Ausf B chassis being used for driver training.

Maximum road speed	40 kph/24.9 mph
Road range	140 km/87 miles
Cross-country range	115 km/71.4 miles
Length overall	4,420 mm/174 in
Width	2,060 mm/81.1 in
Height	1,720 mm/67.7 in
Engine	One 6-cylinder Maybach NL38TR (100 hp)
Track width	280 mm/11 in
Wheel base	1,670 mm/65.75 in
Armament	2 × 7.92 mm MG
Ammunition carried	1,525 rounds
Bow armour	13 mm/0.51 in
Side armour	13 mm/0.51 in
Roof and floor armour	6 mm/0.24 in
Turret armour	13 mm/0.51 in
Crew	2

PzKpfw II (SdKfz 121)

Origins

Development contracts for a projected ten-ton tank were issued in July 1934. Three firms submitted prototypes which were tested rigorously until the contract was given to MAN, and the first vehicles were produced in 1935. These vehicles were used for development only, and all had a 20 mm cannon mounted in the turret with a co-axial machine-gun. As a result of this development, production vehicles had thicker armour and a more powerful engine.

Models

PzKpfw II Ausf A First produced during 1937, this became one of the most widely used vehicles in service during 1939 and 1940. It had an angled front hull.

PzKpfw II Ausf B and C These two models were almost identical and differed from the Ausf A in having a prominent turret cupola.

PzKpfw II Ausf D and E The Ausf D and E were built by Daimler-Benz and differed from other models by having a different Famo/Christie suspension. First produced in 1938, they had larger road wheels but retained the earlier superstructure, and were capable of speeds up to 55 kph. The conversion was not a success as the suspension was too weak for prolonged cross-country work, and the variant was withdrawn in 1940. The chassis were then converted to other uses.

Above A PzKpfw II Ausf A in 1940. **Left** This PzKpfw II Ausf B or C has the markings of the Afrika Korps and the 21st Panzer Division on its glacis plate.

German tanks of World War 2

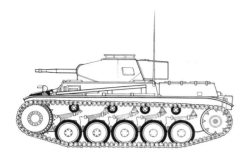

PzKpfw II   1:76 scale

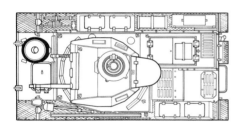

PzKpfw II Ausf F The Ausf F reverted to the earlier suspension of the Ausf A, B and C. It had thicker armour and some detail changes but the armament remained unchanged. A revised cupola was fitted.

PzKpfw II Ausf G and J These two models were almost identical to the Ausf F but had a stowage bin fitted to the back of the turret.

Following on from the above models came a series of vehicles based on the PzKpfw II but featuring heavier armour and revised suspensions with inter-leaved road wheels. These prototypes were not developed beyond the pro-totype stage until one, the VK 1303, was selected as the basis for a light reconnaissance tank which emerged as the Luchs.

PzKpfw II Ausf L (SdKfz 123) This model was named the Luchs (Lynx) and was built in late 1942. It entered service in early 1943. Despite the numerous improvements made to the basic design, the armament remained the 20 mm cannon and one machine-gun, but after 100 had been produced, a further 31 were fitted with a 5 cm gun. Production ceased in May 1943. The Luchs was the last of the German light tanks in production and service, for after 1943 production was switched to the heavier tanks. Exactly why this lightly armed vehicle was kept in production at such a late stage of the war is difficult to determine. Perhaps the answer was that the Luchs was intended as a reconnais-sance tank only, and can be regarded as a tracked armoured car.

Variants

Flammpanzer II Many of the PzKpfw II Ausf D and E vehicles withdrawn from service were converted to flame-thrower tanks by the addition of two flame projectors on each front track

PzKpfw II Ausf L Luchs.

cover. The crew was reduced to two, and the vehicle had a flamethrowing range of about 40 yards. One machine-gun was retained for defence. About 95 were converted.

Geschützwagen II für 15 cm sIG 33 There were two versions of this self-propelled artillery carriage. The first was a simple conversion of an Ausf C to carry the standard infantry heavy support weapon. It entered service in 1942, but it soon became apparent that the chassis was overloaded, and a second version appeared during 1943 on which the chassis was lengthened by the addition of an extra roadwheel.

Geschützwagen II für 7.5 cm Pak 40/2 (Marder II) The Marder II (Martin II) entered service in 1942 and was one of the more successful of the numerous Panzerjäger vehicles. It mounted a special version of the hard-hitting 7.5 cm Pak 40 anti-tank gun, and Ausf A, C, and F chassis were used for the conversion. A total of 1,217 were made, and the type served on many fronts.

PzJäg II Ausf D, E für 7.62 cm Pak

Data PzKpfw II

	Ausf D, E	Ausf F	Luchs
Weight in action	10 tons	9.5 tons	11.8 tons
Maximum road speed	55 kph/34 mph	40 kph/24.8 mph	60 kph/37.3 mph
Road range	200 km/124 miles	200 km/124 miles	250 km/155 miles
Cross-country range	130 km/80.7 miles	100 km/62 miles	150 km/93 miles
Length overall	4,640 mm/182.7 in	4,810 mm/189.3 in	4,630 mm/182 in
Width	2,300 mm/90.5 in	2,280 mm/89.8 in	2,490 mm/98 in
Height	2,020 mm/79.5 in	1,980 mm/78 in	2,130 mm/83.8 in
Engine horse power	140	140	180
Track width	300 mm/11.8 in	300 mm/11.8 in	360 mm/14.2 in
Wheel base	1,920 mm/75.6 in	1,920 mm/75.6 in	2,070 mm/81.5 in
Armament	1×KwK 30 or 38	1×KwK 30 or 38	1×KwK 30
	1×7.92 mm MG	1×7.92 mm MG	1×7.92 mm MG
Ammunition carried	180×20 mm	180×20 mm	330×20 mm
	1,425×7.92 mm	2,550×7.92 mm	2,550×7.92 mm
Bow armour	30 mm/1.2 in	35 mm/1.38 in	30 mm/1.2 in
Side armour	14.5 mm/0.57 in	20 mm/0.79 in	30 mm/1.18 in
Roof armour	14.5 mm/0.57 in	14.5 mm/0.57 in	13 mm/0.51 in
Turret armour (front)	30 mm/1.18 in	30 mm/1.18 in	30 mm/1.18 in
Crew	3	3	4

German tanks of World War 2

36(r) During the early stages of the Russian campaign the T-34 tank was soon found to be invulnerable to most German weapons. As a result large numbers of captured Russian Model 1936 field guns were converted to anti-tank guns and some were mounted on redundant Ausf D and E chassis. These vehicles were rushed into action, despite their open fighting compartments, and were used as tank-hunters.

Geschützwagen II für 10.5 cm 1eFH 18/1 Wespe One of the most successful of all the mobile field artillery pieces produced in Germany during 1939-1945 was the Wespe (Wasp). It was a conversion of the basic PzKpfw II chassis to carry a standard field artillery piece, and the type was produced in large numbers — 683 were in service in 1942. Normal crew was four men. Some were produced minus the gun and were used for carrying ammunition.

Amphibious PzKpfw II Ausf A A small number of vehicles were converted for amphibious warfare in preparation for Operation Seelöwe (Sea Lion) during 1940. Despite successful trials the type was not used in action.

Above right *Geschutzwagen II für 15 cm sIG 33.* **Above left** *Panzerjager II Ausf D or E für 7.62 cm Pak 36(r). This tank-killer conversion used the chassis of the PzKpfw II Ausf D or E which had larger road wheels than other PzKpfw II models and also a different suspension. The gun was a captured Russian field gun converted by the Germans to an anti-tank gun.* **Below** *Geschützwagen II für 10.5 cm leFH 18×1 Wespe*

PzKpfw III (SdKfz 141)

Origins

The vehicle that was to become the PzKpfw III was originally intended to be the main fighting vehicle of the new panzer units but this idea was soon dropped once the early war campaigns had been analysed. The original vehicles were ordered under the cover designation of 'Zugführerwagen' (platoon commander's vehicle) in 1936, and Daimler-Benz and Rheinmetall built prototypes — a Krupp design was built but was not successful. The Daimler vehicle was selected for production and the first was built in 1937.

It is interesting to note that at the design stage a 5 cm gun was considered for the main armament, but a 3.7 cm weapon was considered to be sufficient. However, the turret ring was designed to be large enough for any possible gun that might be fitted — a remarkable example of foresight.

Models

PzKpfw III Ausf A Only ten of this version were built, and they featured rather large road wheels, an internal mantlet and coil spring suspension which was soon shown to be too flimsy for its task. The vehicles were built during 1937.

PzKpfw III Ausf F in the Western Desert, and armed with a 5 cm KwK 39.

PzKpfw III Ausf M complete with Schuetzen and 5 cm KwK L/60 gun.

PzKpfw III Ausf B On this model, the road wheels were smaller and a revised suspension was tried.

PzKpfw III Ausf C Built during 1938, the Ausf C had a revised suspension again, but was otherwise the same as the Ausf B.

PzKpfw III Ausf D This model had increased armour and a revised cupola.

PzKpfw III Ausf E Up till this model the suspension design was never satisfactory and on the Ausf E the problem was cured by installing a six-wheel torsion bar system which was used on later models. A more powerful engine was fitted, and some late examples had a 5 cm gun and an external mantlet. This model formed the bulk of the PzKpfw IIIs in service during the early war campaigns.

PzKpfw III Ausf F Where armament is concerned, the division between the Ausf E and F is not clear for both could be found with either the early 3.7 cm gun or the later 5 cm KwK 39 L/42 gun. The Ausf F production began in late 1940, and the type was the first PzKpfw III with a turret stowage bin, although some were retrofitted to earlier models.

PzKpfw III Ausf G The main change on this model was to the cupola. Some Ausf G vehicles were tropicalised for service in North Africa by the addition of filters for the engine and had the (Tp) designation added to their title.

PzKpfw III Ausf H In 1941 the basic

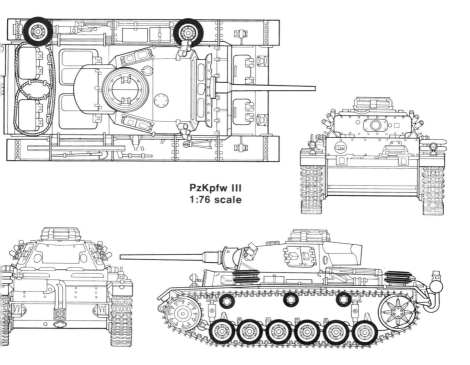

PzKpfw III
1:76 scale

PzKpfw III underwent a major revision with wider tracks, more armour, a new, drive transmission and numerous design changes intended to speed production and cut down the use of scarce raw materials. Some vehicles were fitted with the longer 5 cm L/60 gun.

PzKpfw III Ausf J This model had thicker armour and most were fitted with the 5 cm L/60 gun. The SdKfz number became SdKfz 141/1.

PzKpfw III Ausf L The Ausf K did not see service so the next model was the Ausf L. This had the L/60 gun fitted as standard, and spaced armour was fitted to increase protection. The model was built during 1942 and some were sent to Africa. Many were retrofitted with hanging plate armour.

PzKpfw III Ausf M Built during 1942, this model was further simplified for production, but was essentially the same as the Ausf L.

PzKpfw III Ausf N The Ausf N was a remarkable example of German

ingenuity for it mounted short 7.5 cm guns taken from early models of PzKpfw IV tanks when they were fitted with longer guns. 666 were built and some were converted from Ausf M vehicles. They were used as infantry support vehicles.

The Ausf N was the last PzKpfw III model to be built but plans were made to produce a PzKpfw III/IV model

PzKpfw III Ausf N armed with a short 7.5 cm gun.

Left Stug III with short 7.5 cm gun — this was the first type of Stug III to see service. **Below** *A Stug III Ausf G in its final production form with a Saukopfblende cast gun mounting which replaced the earlier bolted mantlet. This vehicle also is of note in that it is in winter camouflage and uses a two-digit tactical sign 02 on its side.*

incorporating assemblies from both the PzKpfw III and IV production lines. Some self-propelled gun platforms were built but no tanks were produced.

Variants

Most numerous of the Panzerkampfwagen III variants were the Sturmgeschutz III assault guns. Originally armed with the short 7.5 cm Kwk L/24 gun as fitted to the early PzKpfw IV series, the Stug III was later fitted with the L/43 and L/48 7.5 cm guns and the 10.5 cm Sturmhaubitze. There were many different versions of the Stug III with the main differences

being in the superstructure and running gear. The main SdKpfz number for the series overall was SdKfz 142.

StuIG 33 auf Fgst PzKpfw III In 1941 a small batch of 12 PzKpfw III Ausf H vehicles were converted to take the infantry 15 cm sIG 33 gun, housed in a closed box in place of the normal turret. This variant saw service in Russia.

PzKpfw III Flammpanzer III (SdKfz 141/3) In 1942 some Ausf H and M tanks had their main armament replaced by a flamethrower, and were issued to special flamethrower units. Crew was reduced to three men and two machine-guns were retained.

German tanks of World War 2

PzKpfw III (Tauchfähig) (Submersible) This variant was prepared during 1940 for the intended invasion of Britain and used Ausf E or F tanks. The tank was sealed and used a schnorkel while under water for air breathing. They were used in action in the invasion of Russia in 1941.

Panzerbeobachtungswagen III (SdKfz 143) This was a special observation vehicle produced in 1941 for the control of self-propelled artillery fire. The gun was replaced by a dummy barrel offset to enable a machine-gun to fire through the mantlet. The inside of the vehicle was occupied by radios and observation instruments.

Panzerbefehlswagen III There were several types of armoured command vehicles based on the PzKpfw III. They differed from normal tanks only in having extra radio equipment and the attendant aerials. There were several possible radio 'fits' and these were denoted by the SdKfz numbers of 266,267 and 268. Models used were the Ausf D, E, H and K.

Bergepanzerwagen III This turretless variant was equipped as an armoured recovery vehicle.

Schlepper III On the Russian front some elderly PzKpfw III vehicles had their turrets replaced by a flat wooden platform for use as supply vehicles.

This Stug III Ausf G is armed with a 10.5 cm StuH 42 howitzer. On this vehicle the tactical sign 55 is on the barrel sleeve.

Below *This Panzerbeobachtungswagen III picture is interesting for several reasons. First it is in winter camouflage. It is also fitted with Schuetzen and shows clearly how far away from the hull they were fitted. The dummy barrel is clearly shown in its offset position with its normal central location ready for a machine-gun to be mounted.*

German tank data

Leading this column of PzKpfw IIs is a Panzerbefehlswagen III Ausf E, easily identified by the rail-like aerial mounted over the rear hull. Although it is difficult to see the tactical sign on the turret side is IN1.

Data PzKpfw III

	Ausf B	Ausf G	Ausf N
Weight in action	15 tons	20.3 tons	22.3 tons
Maximum road speed	32 kph/19.8 mph	40 kph/24.8 mph	40 kph/24.8 mph
Road range	150 km/93 miles	175 km/108 miles	175 km/108 miles
Cross-country range	95 km/59 miles	97 km/60 miles	97 km/60 miles
Length overall	5,690 mm/224 in	5,410 mm/213 in	5,520 mm/217.3 in
Width	2,810 mm/110.6 in	2,920 mm/115 in	2,950 mm/116.1 in
Height	2,540 mm/100 in	2,440 mm/96 in	2,510 mm/98.8 in
Engine horse power	230	300	300
Track width	360 mm/14.1 in	360 mm/14.1 in	400 mm/15.75 in
Wheel base	2,490 mm/98 in	2,490 mm/98 in	2,510 mm/98.8 in
Armament	1×3.7 cm KwK	1×5 cm KwK	1×7.5 cm KwK
	3×7.92 mm MG	2×7.92 mm MG	2×7.92 mm MG
Ammunition carried	150×3.7 cm	99×5 cm	64/7.5 cm
	4,500×7.92 mm	2,00×7.92 mm	3,450×7.92 mm
Bow armour	14.5 mm/0.57 in	30 mm/1.18 in	50 mm/1.97 in
Side armour	14.5 mm/0.57 in	30 mm/1.18 in	30 mm/1.18 in
Roof armour	18 mm/0.7 in	18 mm/0.7 in	18 mm/0.7 in
Turret armour	14.5 mm/0.57 in	30 mm/1.18 in	57 mm/2.24 in
Crew	5	5	5

PzKpfw IV (SdKfz 161)

Origins

The first orders for a design that was to become the PzKpfw IV were issued in 1934. Three firms built prototypes, but the Krupp submission, which was designated VK2001, was given the production contract in 1936. As things turned out the final production version was a combination of features from both the Krupp and Rheinmetall designs. The project intention was disguised by the appelation of 'battalionsführerswagen' (battalion commander vehicle), and production began in 1937 after extensive trials.

Models

PzKpfw IV Ausf A It is a tribute to the sound basic design of the PzKpfw IV that the first production vehicle was essentially the same basic vehicle as the last tank to roll off the production line. Between 1937 and 1945 the PzKpfw IV was fitted with more powerful guns and thicker armour but the suspension and drive systems remained unchanged. 35 Ausf A tanks were built and all were fitted with the short 7.5 cm gun that made the PzKpfw IV one of the most heavily armed tanks of its day.

PzKpfw IV Ausf B This model differed from the Ausf A in detail only — for example the cupola was revised

Above *PzKpfw IV Ausf B or C.* **Below** *PzKpfw IV Ausf E.*

and the hull front was simplified.

PzKpfw IV Ausf C More small design changes were added to produce the Ausf C, the most prominent of which was the addition of a sleeve to protect the turret machine-gun. The engine was also changed.

PzKpfw IV Ausf D In 1940 the Ausf D was introduced which incorporated for the first time the sloped roof to the hull front roof. Other changes were to

the track and a more powerful engine was installed.

PzKpfw IV Ausf E This model was intended as an interim model only, pending the production of the Ausf F. Extra armour was fitted and many details intended for the Ausf F were incorporated.

PzKpfw IV Ausf F The Ausf F was intended to be the main production variant of the PzKpfw IV but it was

Top *A PzKpfw IV Ausf F2 and its crew at rest 'somewhere in Russia'. The tank in the background is a PzKpfw III Ausf J.* **Above** *PzKpfw IV Aust H in the Russian winter. Items to note here are the streaky winter paint finish and the missing side armour plates on the following vehicle.*

soon overtaken by events as the short L/24 gun was replaxed by an L/43 version. With this change the PzKpfw IV was no longer a support tank for other forces but it became an excellent fighting tank and was used as such from 1941 onwards, superceding the PzKpfw III. Later this version was redesignated the F1 as an even better gun was installed.

PzKpfw IV Ausf F2 The early Ausf F model fitted with the short L/24 gun

German tanks of World War 2

was later retrofitted with the longer L/43 version, and in time with the later L/48 weapon. It was used in large numbers and saw service on all fronts, including the Western Desert.

PzKpfw IV Ausf G This model was basically the same as the Ausf F2 but had thicker armour. The Ausf G was the first model to be fitted with 'Scheutzen' side armour.

PzKpfw IV Ausf H With the introduction of the Ausf H in 1943, the PzKpfw IV took on a new lease of life, for it was fitted with the potent 7.5 cm KwK 40 L/48. With this gun the PzKpfw IV was able to take on almost any tank, and was thus able to retain its placing in the panzer divisions. Changes to help in speeding production were made and 'scheutzen' were standard.

PzKpfw IV Ausf J The Ausf J was the last production model and appeared in 1944. By that time many raw materials and processes were difficult to obtain, and this was visible in the use of wire mesh 'scheutzen' in place of the usual steel plates. Many other changes had to be made to simplify the design but it remained basically unchanged from earlier models.

Variants

As the PzKpfw IV was produced in larger numbers than any other German tank it is not suprising that it was much used for many other tasks apart from that of battle tank. The listing below can only mention the more common tasks that the PzKpfw IV chassis had to perform.

Jagdpanther IV with the 7.5 cm L/70 gun.

Below *Stug IV column. The driver's armoured box is clearly visible.*

Jagdpanzer IV Ausf F One of the most successful of all the PzKpfw IV variants was the Jagdpanzer IV. This was a conversion of a standard PzKpfw IV chassis to take a low fighting compartment with well-sloped armour. Armament was the 7.5 cm StuK L/48. This version had the SdKfz number 162 but a later variant with an L/70 gun became the SdKfz 162/1. This later variant was more cumbersome than the version with the L/48 gun and was more difficult to handle, but it was pressed into service as 'it was the Führer's will'. A later variant still was the 'Zwischenlosung' which was the addition of a Jagdpanzer IV superstructure mounted direct onto a PzKpfw IV chassis. In all its forms the Jagdpanzer IV was a formidable opponent and an effective tank-killer.

Sturmgeschutz IV (SdKfz 163) In 1943 some spare capacity was found which was able to turn out PzKpfw IV chassis, and this was utilised for a time in a strange conversion in which StuG III superstructures were added to the PzKpfw IV chassis. The result was the StuG IV which served alongside the StuG III. Many of these vehicles carried extra concrete armour.

Sturmpanzer IV Brummbär (Grizzly Bear) The German experiences in such cities as Stalingrad and Leningrad convinced the Germans that they needed a specialised vehicle for street fighting. They added a heavily armoured compartment to a PzKpfw III chassis and armed it with a 15 cm L/12 gun — the result was the Brummbär. Ausf F, G, H and J vehicles were used and the Brummbär was produced in some numbers with several variations.

Panzerjäger III/IV Nashorn or Hornisse (Rhinocerous or Hornet) (SdKfz 164) In 1942 the need for a heavy anti-tank gun was desperate and a typical German improvisation emerged in the shape of a PzKpfw IV chassis with PzKpfw III drives, mounting an 8.8 cm Pak 43/1 gun. The result was high and rather awkward but it worked and was produced in some numbers pending better equipment.

Flakpanzer IV There were several variations of Flakpanzer IVs. One

Above *The two forms of turreted Flakpanzers built on the PzKpfw IV chassis. In the foreground is the 3.7 cm Flak 43 auf Sf Ostwind, and in the background a Flakpanzer IV (20 mm) auf Fgst Pz IV/3 Wirbelwind.* **Below** *A 3.7 cm Flak 43 auf Sf IV Mobelwagen.*

Sturmpanzer IV Brummbär covered with a Zimmerit coating.

German tanks of World War 2

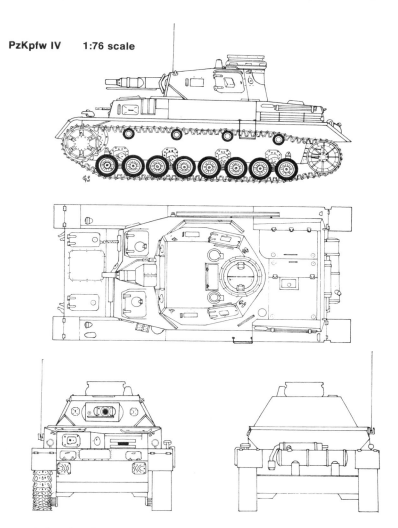

PzKpfw IV   1:76 scale

mounted a four-barrel 20 mm Flak-vierling 38 on an open platform, another the same gun in an enclosed turret, and two others mounted a single 3.7 cm Flak 43 gun — again, one on an open platform and the other in a turret. These vehicles, known as the Mobelwagen (Furniture Van), Ost-wind (East Wind) and Wirbelwind (Whirlwind), were used to give some form of anti-aircraft defence to armoured units.

Geschützwagen III/IV Hummel (Bumble Bee) (SdKfz 165) This was another PzKpfw III/IV vehicle, this time used to mount the 15 cm schwere Panzerfeldhaubitze 18/1 field piece. First produced in 1942, the Hummel was manufactured in large numbers.

On the Russian front many old PzKpfw IV tanks had their turrets removed and replaced by a flat plat-form or truck platform. They were then used as supply vehicles. Some were used as engineer vehicles carrying bridging equipment, and others were used as armoured recovery vehicles.

German tank data

Data PzKpfw IV

	Ausf C	Ausf F2	Ausf J
Weight in action	20 tons	23.6 tons	25 tons
Maximum road speed	40 kph/24.8 mph	40 kph/24.8 mph	38 kph/23.6 mph
Road range	200 km/124 miles	200 km/124 miles	300 km/186 miles
Cross-country range	130 km/80.7 miles	130 km/80.7 miles	180 km/111.8 miles
Length overall	5,870 mm/231 in	6,630 mm/261 in	7,020 mm/276 in
Width	2,850 mm/112 in	2,880 mm/113.4 in	3,290 mm (with skirts) 129.5 in
Height	2,590 mm/102 in	2,680 mm/105.5 in	2,680 mm/105.5 in
Engine horse power	300	300	300
Track width	380 mm/15 in	400 mm/15.75 in	560 mm (Ostketten) 22 in
Wheel base	2,620 mm/103 in	2,620 mm/103 in	2,620 mm/103 in
Armament	1×7.5 cm L/24	1×7.5 cm L/43 or L/48	1×7.5 cm L/48
	2×7.92 mm MG	2×7.92 mm MG	2×7.92 mm MG
Ammunition carried	80×7.5 cm	87×7.5 cm	87×7.5 cm
	2,700×7.92 mm	3,150×7.92 mm	3,150×7.92 mm
Bow armour	30 mm/1.18 in	30+30 mm/ 1.18+1.18 in	80 mm/3.15 in
Side armour	14.5 mm/0.57 in	20+20 mm/ 0.78+0.78 in	30+5 mm/ 1.18+0.2 in
Roof armour	11 mm/0.43 in	11 mm/0.43 in	16 mm/0.63 in
Turret armour	30 mm/1.18 in	50 mm/1.97 in	50 mm/1.97 in
Crew	5	5	5

PzKpfw V Panther (SdKfz 171)

Origins

Soon after the invasion of Russia in 1941, the panzer troops encountered the Russian T-34 tank. The T-34 was well armed with a 76.2 mm gun, had well sloped and effective armour, and was fast and handy. It outfought all the German tanks then in service with the exception of the PzKpfw IV, so its appearance had a profound effect on the German panzer arm. The T-34 design was closely studied and at one time it was proposed that it should be copied direct and produced in Germany. Troops at the front went further and pressed large numbers of captured T-34s into service against their former owners, but the German national pride could not accept a direct copy.

Instead, the main features of the T-34 were incorporated into a new German design which became the Panther. A design proposed by MAN was accepted in September 1942 and the production of the new tank was given the highest priority — the first tank came off the line in November 1942. This was the Ausf D1 (Panther ausf numbers did not run in sequence), which was soon followed by the Ausf D2, the production variant.

The Panther eventually became the best of all the German tanks, but its baptism of fire was a disaster for it was pressed into service during the Battle of Kursk (which was delayed in order to allow the Panther to participate) at a time when it was not fully developed or tried, and breakdowns were frequent. After this early misadventure the Panther became an excellent fighting tank.

Models

Panther Ausf D2 The Panther had well-sloped armour, a powerful 7.5 cm L/70 gun and interleaved road wheels supported on torsion bar suspension. It was fast for its size but the final product turned out to be overweight and was thus not so handy as had been hoped. When the first Panthers went into action many defects, both design and mechanical, were discovered the hard way, not the least of which was

Above *Panther Ausf D2. This model can be recognised by the letter-box machine-gun slot on the glacis plate.* **Below** *Panther Ausf A. On this model the machine-gun mounting was changed to a ball mount and the cupola shape was altered.*

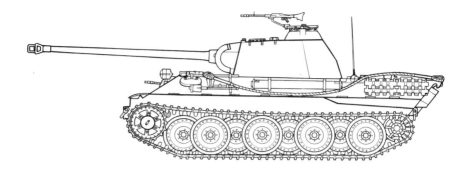

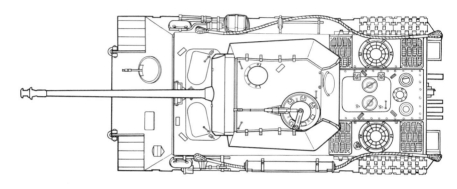

**PzKpfw V
Panther
1:76 scale**

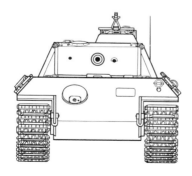

German tanks of World War 2

Panther Ausf G. This was the last Panther model to be produced although others were planned. The driver's hatch has been removed and replaced by periscopes while the hull side armour has been altered. The tube seen on the hull side was for gun barrel cleaning rods.

that the fuel tanks in the hull rear were insufficiently armoured and caught fire easily. Many of these defects were eliminated in later models.

Panther Ausf A The next Panther model was the Ausf A which was first produced in late 1943. A more powerful engine was fitted and the hull

machine-gun port was changed from a slot to a ball mounting. Numerous other changes were made and 'scheutzen' side armour was added.

Panther Ausf G This was the final production model but it existed in more than one version. The main change from the earlier models was

Data PzKpfw V

	Ausf D	Ausf A	Ausf G
Weight in action	43 tons	45.5 tons	44.8 tons
Maximum road speed	46 kph/28.6 mph	46 kph/28.6 mph	46 kph/28.6 mph
Road range	169 km/105 miles	177 km/110 miles	177 km/110 miles
Cross-country range	85 km/52.8 miles	89 km/55.3 miles	89 km/55.3 miles
Length overall	8,860 mm/348.8 in	8,860 mm/348.8 in	8,860 mm/348.8 in
Width	3,430 mm/135 in	3,430 mm/135 in	3,430 mm/135 in
Height	2,950 mm/116 in	3,100 mm/122 in	3,000 mm/118 in
Engine horse power	650	700	700
Track width	650 mm/25.6 in	650 mm/25.6 in	650 mm/25.6 in
Wheel base	2,620 mm/103 in	2,620 mm/103 in	2,620 mm/103 in
Armament	1×7.5 cm L/70	1×7.5 cm L/70	1×7.5 cm L/70
	2×7.92 mm MG	3×7.92 mm MG	3×7.92 mm MG
Ammunition carried	79×7.5 cm	79×7.5 cm	82×7.5 cm
	4,104×7.92 mm	4,200×7.92 mm	4,200×7.92 mm
Bow armour	80 mm/3.15 in	80 mm/3.15 in	80 mm/3.15 in
Side armour	40 mm/1.57 in	40 mm/1.57 in	50 mm/1.97 in
Roof armour	15 mm/0.59 in	15 mm/0.59 in	40 mm/1.57 in
Turret armour	120 mm/4.7 in	120 mm/4.7 in	120 mm/4.7 in
Crew	5	5	5

German tank data

Above *Jagdpanther.* **Below** *Bergepanzer Panther.*

that the hull shape was revised to give more armour protection and also to make it easier to manufacture. The driver's vision port in the hull front was removed and replaced by a peri-scopic vision device, and many other changes were incorporated. The first vehicles off the line in 1944 continued to use the dished and convex inter-leaved roadwheels, but on the later

German tanks of World War 2

versions these were replaced by the steel wheels used on the Tiger suspension in its late production form. The Panther was still in production as the war ended.

If the war had continued beyond May 1945, it had been proposed that a Panther II would replace the earlier models in production. This model would have used a smaller turret which would give more protection and which was capable of mounting an 8.8 cm gun.

Variants

Jagdpanther (SdKfz 173) Many armour experts regard the Jagdpanther as one of the best armoured fighting vehicles to emerge from World War 2. It was a conversion of the basic Panther chassis to take a well-shaped sloping superstructure which could mount the very effective 8.8 cm Pak 43/3. This vehicle could outrange and outfight nearly every Allied tank it was likely to encounter, and in addition it was fast and handy. Three hundred and eighty-two were built during 1944 and 1945, and they were respected opponents.

Bergepanzer 'Panther' (SdKfz 179) When such heavy tanks as the Tiger and Panther entered service, existing armoured recovery vehicles were not capable of assisting and recovering such vehicles from the battlefield. The solution was to convert some of the older Ausf D and A vehicles by removing the turret and fitting heavy-duty winches inside the hull. As the Bergepanzer 'Panther', 297 vehicles were converted.

Beobachtungspanzer Panther (SdKfz 267 or 268) A small number of Panthers were converted to the artillery observation role by removing the main gun and replacing it with a false barrel. The inside of the turret was then equipped for its special task, and a machine-gun was fitted on one side of the turret front, Extra radios and vision devices were added.

As well as the above variants, planned vehicles were to be Flak tanks, mineclearing vehicles, and one project was for a tank-killer mounting a

12.8 cm gun. One project intended that a shortened Panther chassis was to carry a 10.5 cm field piece in a turret that could be emplaced as a form of pill-box emplacement, and recovered by the same vehicle when required. If this varient had been built it would have been a classic example of German inability to concentrate their efforts on producing large numbers of successful vehicles, for such a project would only have diverted much-needed design and production facilities, to say nothing of raw materials.

PzKpfw VI Tiger (SdKfz 181)
Origins

Up till 1941 it was felt that the PzKpfw III and IV were adequate for any tasks likely to be encountered, but this complacency had already been ruffled in France in 1940 when it was discovered that many French and British tanks, especially the British Matilda, were more heavily armoured than their German counterparts.

The discovery of the considerable combat potential of the Russian T-34 and KV-I in 1941 therefore showed the German designs to be at a disadvantage. Hitler himself had envisaged the need for a new and heavier tank design in May 1941 and the events in Russia seemed to confirm the accuracy of his 'intuition'.

The result was a design specification for a heavy tank mounting a 8.8 cm gun and having sufficient armour to defeat all the likely future anti-tank weapons. Two firms, Porsche and Henschel, submitted designs for what was given the design designation of VK 3601, and both firms built prototypes mounting a large Krupp-designed turret. The Porsche design had many novel features including a petrol-electric drive system, but it was not selected for service.

The chassis was, however, selected as a heavy tank-killer mobile gun carrier, and eventually emerged as the Ferdinand or Elefant. In this form it became one of the most monumental failures of the German armament

Jagdpanzer Tiger(P) Elefant mit 8.8 cm Pak 43/2. This tank destroyer was built on to the hull of the projected Porsche Tiger.

industry at the Battle of Kursk, for it was underdeveloped and lacked a secondary defensive machine-gun.

But to return to the Henschel design — this was chosen for production and was named the PzKpfw VI or Tiger. Production began slowly in August 1942 but soon increased in volume after the personal intervention of Hitler.

At the time of its introduction, the Tiger was the most powerful tank in the world. It was armed with the formidable 8.8 cm KwK 36 which had been developed from the 8.8 cm Flak 18 and 36 anti-aircraft gun. At its thickest, the Tiger's armour was a hefty 102 mm and was thus almost invulnerable to all anti-tank weapons then in use.

But its main disadvantage was its bulk and weight. Its bulk was such that it was too heavy for most European bridges and had to be fitted with wading equipment and air 'schnorkels'. It was too wide for most railway flatcars and had to have two tracks — one for action and another narrower set for railway transport, which also involved removing roadwheels and side shields. Weight was 56 tons which severely restricted its battlefield hand-

ling, but the Tiger was considered to be a formidable opponent and Allied armies had to evolve special tactics to counter the Tiger. Production ceased in August 1944 by which time 1,350 Tigers had been delivered.

Models

PzKpfw VI Tiger Ausf E Only one Tiger model, the Ausf E, was built but it was made in several different versions. The first version had extensive wading equipment and triple overlapping road wheels. These road wheels were arranged in three rows and were suspended on torsion bars. The outer road wheels had to be removed for railway transport and narrower tracks were then used which involved a considerable expenditure of time and trouble. Another unforeseen disadvantage of the interleaved road wheel system was discovered in Russia when mud, slush and snow froze solid between the wheels and immobilised the vehicles.

Late variants used a much simplified road wheel system with steel resilient wheels in place of the earlier solid dished wheels with their rubber tyres. Other changes made included a more powerful engine increased in size

German tanks of World War 2

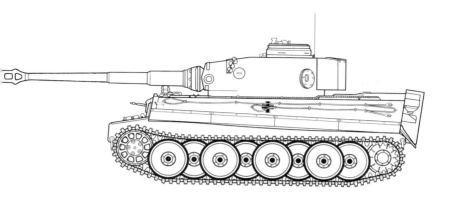

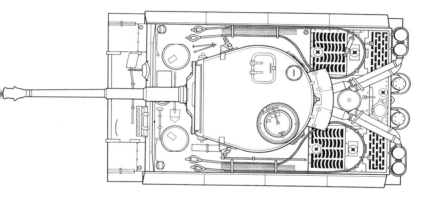

PzKpfw VI Tiger I   1:76 scale

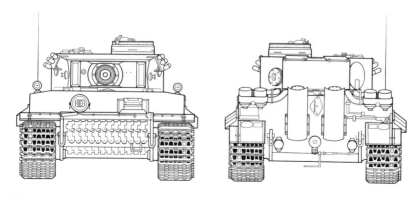

German tank data

from 21 to 24 litres. Later versions also omitted the wading equipment. Tigers built for use in Africa had a special Tp designation denoting suitable for tropical use which involved special air filters for the engine. These filters were also found on Tigers used in Russia.

Other changes incorporated were to the commander's cupola and minor changes to the drive mechanism, but the basic Tiger design remained substantially unchanged to the end. It was the most powerful tank of its day, but with the arrival of the Russian heavy

Below *A brand new Tiger in factory finish. The outer road wheels have been removed for transport purposes and a special narrow width track fitted.* **Bottom** *A Tiger Ausf E in full battle trim with battle tracks.*

German tanks of World War 2

Above A late production Tiger with steel resilient wheels. **Below** American troops examine a captured Sturmtiger. The rocket they are examining in front of the vehicle has been fired but its large size can be appreciated. The large calibre rocket launcher has a counterweight around the muzzle but this was not fitted to all vehicles.

German tank data

tanks that day soon passed, and the Tiger's bulk and weight made it more of a mobile pill-box rather than a fighting tank.

Variants

Panzerbefehlswagen Tiger Ausf E (SdKfz 267 and 268) Some Tigers were converted to command tanks by the addition of extra radio equipment. The two SdKfz variants differed only in the types of radio fitted.

Bergepanzer Tiger A small number of Tigers were converted to the armoured recovery role by having their guns removed and replaced by a winch. The exact number converted was very small and may even have been only one.

Sturmtiger Ten Tiger chassis were converted in 1944 for street fighting by the addition of a large armoured superstructure mounting a 38 cm rocket projector. This weapon was originally intended as a naval anti-submarine device and fired a hefty 345.5 kg (761 lb) rocket intended for house demolition. The conversion was not a success on account of the vehicle's weight (70 tons) and awkward bulk. This variant consumed two gallons of fuel for every mile travelled.

Data PzKpfw VI Tiger I

Weight in action	56 tons
Maximum road speed	38 kph/23.6 mph
Road range	100 km/62 miles
Cross-country range	60 km/37 miles
Length overall	8,240 mm/324.4 in
Width	3,730 mm/146.5 in
Height	2,860 mm/112.6 in
Engine horse power	700
Track width	725 mm/28.5 in
Wheel base	2,830 mm/111.4 in
Armament	1×8.8 cm KwK 36 2 or 3 7.92 mm MG
Ammunition carried	92×8.8 cm 3,920×7.92 mm
Bow armour	100 mm/3.93 in
Side armour	

(top)	80 mm/3.15 in
Roof armour	26 mm/1.02 in
Turret armour	100 mm/3.93 in
Crew	5

Panzerkampfwagen VI Tiger II (SdKfz 182)

Origins

The Tiger had hardly entered service before consideration was being given to its successor. Again, Porsche and Henschel were given development contracts, and the Porsche submission was at first considered to be the most likely contender as it drew heavily on experience gained from the first Tiger development programme.

But the Porsche design again depended on a petrol-electric drive which would depend on the availability of large amounts of copper for the motors and other electrical components. By 1943 copper was in very short supply in Germany so the Porsche design was dropped in favour of the Henschel submission, the VK 4503(H).

By the time that this decision had been made, Porsche turrets were already in production, and about 50 had been made. These turrets were therefore used on the first Henschel-designed chassis. The Henschel design became known as the Tiger II, or Königstiger, and to the Allies it was known as the King or Royal Tiger. It was designed to use as many Panther components as possible, and by the end of the war 484 had been built, with the first production models appearing during early 1944.

Models

PzKpfw VI Tiger II Ausf B Only one model of the Tiger II was built, the Ausf B. It was the heaviest tank to see operational service during World War 2, and also one of the most powerful. Its main armament was the 8.8 cm KwK 43, developed from the 8.8 cm Pak 43 anti-tank gun. At its thickest point, the Tiger II armour was 185 mm

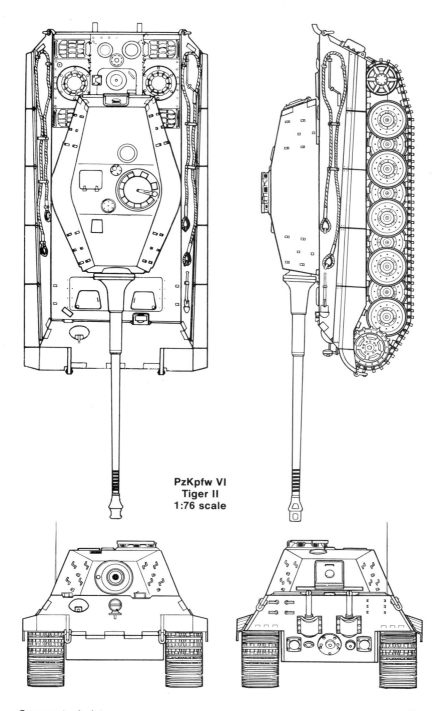

**PzKpfw VI
Tiger II
1:76 scale**

German tank data

Above *PzKpfw VI Tiger II Ausf B fitted with the Porsche turret on the ranges.*
Below *Tiger II with the Henschel turret.*

thick (on the gun mantlet), and the gun and armour went a long way towards the Tiger IIs prodigious weight of 69.7 tons. As the Tiger II used the same engine as the Panther tank it can be seen that it was seriously underpowered, and so performance was severely restricted. Also the Tiger II was rushed into action while still undeveloped and suffered from a long string of mechanical breakdowns and troubles. The first 50 tanks were fitted with the Porsche turret, but the rest had the Henschel tur-

ret which was not only simpler to make but also afforded more protection.

In action, the Tiger II was a formidable opponent which could outshoot and outrange nearly all Allied tanks with the possible exception of the Russian Joseph Stalin series, but its huge weight and size made it ponderous and difficult to conceal. In a swift armoured battle it would have been almost useless but by 1944 Germany was fighting a defensive war and the Tiger II was perfect for that role.

German tanks of World War 2

Jagtiger B. The 12.8 cm gun used on this vehicle fired an HE shell weighing 26 kg (57 lb) and the AP projectile weighed 28.3 kg (62.26 lb). Note the Zimmerit anti-magnetic mine paste which has been liberally applied to the vehicle's hull in an effort to reduce its vulnerability to infantry — a major problem with all self-propelled guns of this type.

German tank data

Variants

Jagdtiger B The most heavily armed of all the German AFVs to see service was the mighty Jagdtiger which mounted a massive 12.8 cm L/55 gun in a superstructure built on to a Tiger II chassis. It was heavily armoured (the front mantlet was 250 mm thick) and almost invulnerable to all opponents, but suffered from the same lack of mobility as the Tiger II. Only 48 had been built, some with a revised Porsche suspension, by the time the war ended.

Data PzKpfw VI Tiger II

Weight in action	69.7 tons
Maximum road speed	38 kph/23.6 mph
Road range	110 km/68.3 miles
Cross-country range	85 km/52.8 miles
Length overall	10,260 mm/403.9 in
Width	3,750 mm/147.6 in
Height	3,090 mm/121.6 in
Engine horse power	700
Track width	800 mm/31.5 in
Wheel base	2,790 mm/109.8 in
Armament	1×8.8 cm KwK 43
	3×7.92 mm MG
Ammunition carried	84×8.8 cm
	5,850×7.92 mm
Bow armour	100 mm/3.93 in
Side armour	80 mm/3.15 in
Roof armour	40 mm/1.57 in
Turret armour (front)	185 mm/7.28 in
Crew	5

five

German tank armament

As a general rule, German tank armament was always well in advance of Allied developments. At a time when British tanks were armed with the 2 pr gun, German tanks were being armed with 5 cm and 7.5 cm guns firing 4.56- and 14-pound projectiles respectively. But tank guns are not rated on the weight of shot that they fire alone.

Another very important factor is the speed at which the shot or shell (shot is solid metal, and shell contains a high explosive charge) leaves the muzzle of the gun. This is usually referred to as the muzzle velocity or V_0.

Here again, the Germans were usually well in advance of Allied gun designers, and in the two 8.8 cm guns that were produced for tanks, they were able to combine a high muzzle velocity and shot weight that could outrange and destroy almost any Allied counterpart.

The most usual way to increase the muzzle velocity of a gun is to increase the charge in the propelling cartridge, but this places severe strain on the breech mechanism and recoil system. Another method, often employed by German designers, was to increase the length of the barrel. The 7.5 cm gun used on the PzKpfw IV was increased from L/24 to L/43 and eventually to L/48, and each increase in length brought about an increase in the hitting power of the gun.

But an increase in barrel length also brought about severe strain on the recoil mechanism, which was usually minimised by the use of a muzzle brake which could reduce recoil forces by directing a proportion of the muzzle blast to the rear at the time the shell left the barrel.

Starting at the bottom of the scale, all German tanks mounted at least one machine-gun. The standard German calibre for machine-guns was 7.92 mm (0.312 in). The most widely used machine-gun fitted to tanks was the MG34, a Rheinmetall weapon which was at one time the standard German infantry machine-gun. On tanks the MG34 was mounted co-axially with the main armament and was often carried in a complex ball-mounting on the front glacis plate. By the time the war ended most German tanks had an extra machine-gun fitted on the commander's cupola for anti-aircraft use.

After about 1943, a new gun known as the MG42 began to be issued. This gun was cheaper and simpler to produce than the MG34, and was designed to replace the earlier gun, but events were such that production could not keep pace with demand and the MG34 was still in use in 1945 as the war ended.

Going up the calibre scale, the next tank gun was the 20 mm KwK 30. This was a shortened version of the 20 mm Flak 30 anti-aircraft gun. This weapon was originally a Swiss Solothurn design known as the T5-150 or ST 9, and was developed under Rheinmetall guidance. It was used only on the PzKpfw II and on some armoured cars. It was later supplemented by the Mauser 20 mm KwK 38, which was again a shortened version of an anti-aircraft gun, in this case the 20 mm Flak 38. This gun used magazines holding either ten or 20 rounds.

The 3.7 cm gun used on the early PzKpfw III was an L/45 weapon developed from the 3.7 cm Pak 35/36 anti-tank gun. It was not produced in large numbers after 1940.

Next in size came the 5 cm KwK L/42 which was fitted to the early PzKpfw III Ausf E to H. This gun was well in advance of similar tank armament at the time of its introduction to service, but was soon outclassed when it encountered the thick hide of the Russian T-34 and KV-I. This had already been anticipated by Hitler, who had ordered an increase in barrel length to L/60, but production difficulties had prevented this order being carried out. Eventually the L/60 gun was placed in production as the 5 cm KwK 39, and in performance was equal to the 5 cm Pak 38 anti-tank gun in use with Wehrmacht anti-tank units.

The 7.5 cm KwK as originally fitted to the early models of the PzKpfw IV was a low velocity gun firing HE shells. In its original form it was a most effective gun when fired against contemporary tanks, but by the end of 1940 it was realised that it had become ineffective against the heavy armour of some Allied vehicles. It was replaced in service by the interim 7.5 cm KwK 40 with an L/43 barrel. The old L/24 guns that were replaced were not scrapped but were placed in store, only to be withdrawn for use in the late models of the PzKpfw III, and some were also placed on half-tracks for close support duties.

The L/43 guns were soon replaced by the 7.5 cm KwK 40 with an L/48 barrel. This gun was a development of the famous Rheinmetall 7.5 cm Pak 40 anti-tank gun and it was to prove itself one of the most potent and effective of all the German tank guns. It was certainly the most widely used for it was placed on a wide variety of tank destroyer chassis, and one version ended up as an anti-tank gun mounted on aircraft (the 5 cm gun also went airborne).

A further development of the 7.5 cm KwK 40 was the KwK 42 which had an L/70 barrel. This gun was used on the Panther and was a considerable advance as a tank gun on the KwK 40. Plans were in hand to increase this weapon to an L/100 version but this project was dropped before the war ended.

The first 8.8 cm tank gun was the KwK 36 which was fitted to the Tiger. It was a powerful weapon with a considerable range but by 1944 it was considered obsolescent and was replaced in production by the even more powerful 8.8 cm KwK 43. The two guns were not related as the KwK 43 had a larger and more powerful propelling charge, and the KwK 36 was an L/56 gun while the KwK 43 had a barrel length of 71 calibres. In its latter form, the KwK 43 can be considered to have been the best all-round tank gun in use by any of the combatants involved in World War 2.

As the war ended, plans were well advanced on even heavier tank armament. The 12.8 cm gun was intended for use in tank turrets, but under development were smooth-bore guns firing finned projectiles, a High-Low pressure gun, and various rocket-propelled projectiles which were intended to have the tank-killing capacity of current anti-tank guns without the weight or cost penalties involved. As the war ended, none of them were sufficiently advanced for service use.

Tank gun data

	Length	Shot weight (kilograms)	Muzzle velocity (metres/sec)
20 mm KwK 30	L/55	0.119	800-990
20 mm KwK 38	L/55	0.119	800-990
3.7 cm KwK	L/45	0.68	762
5 cm KwK	L/42	2.25	685
5 cm KwK 39	L/60	2.25	1198
7.5 cm KwK	L/24	6.36	385
7.5 cm KwK 40	L/43	6.36	7.40
7.5 cm KwK 40	L/48	6.8	930
7.5 cm KwK 42	L/70	6.8	1120
8.8 cm KwK 36	L/56	7.27	810
8.8 cm KwK 43	L/71	7.27	1000

German tanks of World War 2

six

Camouflage and markings

When one starts to write about the camouflage and markings used on German tanks, one is immediately struck by the number of exceptions to the rule that were used in practice. Although definite guidelines were set out for the manner and type of markings and camouflage to be used in the field, in practice these rules seem to have been either interpreted in differing fashions, or quite simply ignored or amended to suit local tastes or conditions. This was particularly true during the latter days of the war but other exceptions to the rule can be discovered even during the early days of 1939 and 1940. Therefore this section does not set out to describe what

was a rigid marking system universally applied, but merely to give a general outline of what was intended to be applied to tanks in the way af camouflage, tactical markings and insignia. To every rule stated below there were many variations and exceptions, and it should be borne in mind that many of the markings were applied by soldiers with varying degrees of skill, and that they were applied under a wide range of circumstances.

Tank camouflage

German tanks were originally delivered in a dark grey finish known as Panzergrau. This grey shade, which incorporated a blue element, was one of the most widely used and encountered of all German tank camouflage finishes, but after about 1942 or 1943 German tanks were delivered in an alternative sand-coloured shade which had originally been applied to Afrika Korps vehicles serving in Africa. This sand-yellow shade was also found to be very suitable for the conditions on the Russian steppes, but in 1944 tanks were again being delivered in the original grey finish. On occasion, tanks were encountered with a basic earth brown scheme, but this

A typical German camouflage scheme is seen on these two abandoned PzKpfw IV Ausf H tanks in Russia. The colours appear to be olive green drab over a basic sand yellow.

was in the early years of 1939 and 1940 only.

Tanks were rarely used in action in their basic colour schemes alone. In addition to the basic grey or sand yellow, tank units were issued with supplies of paint in shades of olive green, light grey, red-brown and dark yellow. These were applied by the tank crews themselves to suit local conditions, and were applied in a wide, combination of camouflage schemes and colours. The extra colours were either applied by brush of were sprayed on to the tanks using spray equipment issued at company level. On occasion there was no time to apply carefully worked-out schemes

Top row a *Tactical Numbers 1937-1940. At first these numbers were painted on to a small plate of the shape shown, and displayed alongside the driving compartment on the hull. 4 meant Fourth Company, 1 the 1st Platoon, and 7 the tank number in that platoon. Colours white on black.* **b** *Tactical numbers 1940-1945. From 1940 onwards the tactical numbers began to be painted on the turret sides and rear. To identify unit commanders the letters R (regiment), I (1st Battalion), II (IInd Battalion) were*

used, suffixed by 0 and another number, eg 01 for commanding officer. Colours used were red, yellow and white. **Second row** *Divisional Tactical Markings 1940-1945. These symbols identified the organic unit within the division, and also its primary mobility. Rarely used on tanks, but appeared on most other vehicles. Shown are* **c** *Regimental staff,* **d** *motorised infantry,* **e** *armoured infantry and* **f** *motorised reconnaissance units.* **Third row g** *motorised anti-tank (Panzerjäger) unit. A fully-tracked unit would have the appropriate 'track' symbol instead of the wheels. In use 1940-1943.* **h** *motorised engineers,* **i** *motorised signals and* **j** *motorised support unit. In this case the S stood for sanitäts or medical.* **Fourth row k** *towed (motorised) artillery,* **l** *towed (motorised) anti-tank, 1942-1943,* **m** *self-propelled artillery and* **n** *self-propelled (half-track) anti-aircraft units. All these symbols were usually accompanied by a number or letter designating the unit identification — Roman numerals the battalion and Arabic numerals the company. Higher formations and Divisional units mostly omitted this identification. The colour of all Divisional or tactical symbols was yellow or (less common) white. Black was used on a lightly coloured background.*

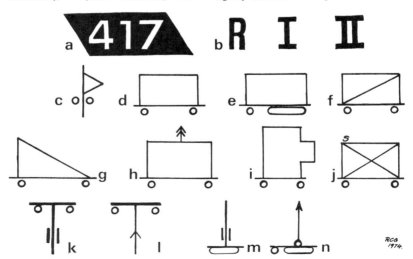

German tanks of World War 2

Another green/sand scheme applied to a 10.5 cm le FH 18 auf Sf 39H(f). This was a standard Wehrmacht field gun mounted on the chassis of a captured French Hotchkiss H39 tank. These conversions were used mainly in France.

and a suitable colour was often applied by throwing the paint at the side of the tank from cans!

The colour combinations were many and varied. A common scheme was sand yellow overlaid by olive green patches or stripes. Some units used schemes which involved the use of the basic grey with red-brown, yellow and green patches. Schemes used were sometimes very elaborate when a unit had time to apply them. Some schemes used splinter or lozenge markings overall, a typical example being the yellow 'spotted' panzer grey overall finish employed in the pine forests of North-West Europe in 1944 and 1945. In the Normandy bocage some Panther units employed a complex lozenge scheme with red-brown and dark grey being the predominant colours.

In winter conditions, when snow covered the ground, German crews followed the usual pattern of daubing their tanks with white paint — not making their vehicles pure white but leaving enough of the basic colour to form a camouflage pattern. White-wash was often used for winter schemes in preference to paint as it could be quickly and easily removed when the snow melted.

These finishes were applied over the basic colour scheme of the tank as delivered from the factory, and from early 1944 onwards tanks were delivered coated in a grey coat of 'Zimmerit'. This was a plaster-like substance applied over the whole of the tank which gave the surface a degree of protection from magnetic anti-tank devices likely to be applied by infantry tank-killer squads. Zimmerit was applied in a rough and corrugated finish which gave the vehicles to which it was applied a matt and worn appearance.

National markings

Perhaps the most universally applied markings used on German tanks was the tactical national recognition marking. This was usually a black cross outlined in white, and was applied to the vehicle sides and rear. Some were also applied to the sides of the turret. In its initial form during 1939 and early 1940 this cross was all white, and the black centre was added

in 1940. There were numerous variations on this theme. On some sand
yellow tanks, the cross was merely
outlined in white or black with the
centre left in the basic sand yellow.
There were also many variations in
shape and size, as well as the actual
positioning on the vehicle.

During the early war years, before
Allied air supremacy became over-
whelming, many German tanks draped
the German national flag over the tank
hull top as a recognition signal for
Luftwaffe aircraft.

As a general rule, captured tanks
used on the same front as they were
captured tended to have the German
crosses very prominently marked in
larger sizes than normal. An obvious
example was the use of T-34 tanks
which were covered with German
crosses.

Top *Preparing for the winter campaign. Applying whitewash with a
crude brush on to a GW II für 7.5 cm
Pak 40/2 (Marder II). This vehicle was
one of the many built on to the chassis
of the PzKpfw II) chassis.* **Above** *A
PzKpfw III Ausf J in winter markings.*

Below *Clearly seen on the front of this
PzKpfw I Ausf B is the early form of
national cross marking used during
the Polish campaign. The vehicles
behind are PzKpfw II tanks and a
SdKfz 251/3 command half-track with
an officer who appears to be none
other than Guderian himself.*

German tanks of World War 2

This picture of a knocked-out Valentine that had been captured and used by the Germans is of interest on several points. First the crosses are white outlined by black instead of the other way around — this was probably a local (North Africa) device for recognition purposes. The Buffalo badge was used by the panzer regiments attached to the 10th Panzer Division, and was usually red. Just visible on the front plate next to the driver's hatch is the symbol of the 10th Panzer Division.

An example of a tank with regimental tactical numbers. This PzKpfw II Ausf B captured in North Africa is coded R06 on the turret side. The Buffalo badge denotes one of the panzer regiments attached to the 10th Panzer Division.

Camouflage and markings   55

When the Germans had air supremacy, such as in France 1940 when this picture was taken, tanks advertised their position to aircraft by using the national flag draped over the rear hull. These tanks are PzKpfw 35(t)s of the 6th Panzer Division. Tne tank in the middle of the picture has the tactical letters 141 on the rear hull plate.

Tactical signs

Most German tank units used an internal recognition scheme based on the use of a three-digit number painted on to the turret sides. This number gave the tank regiment, platoon (zug) number, and the individual tank number in the platoon. For example, a number of 521 would mean that the tank belonged to the fifth regiment, was from the second platoon, and was the platoon leader's tank, as the number one was always reserved for the platoon leader. The second tank in the platoon would be 522 and so on. Regimental tanks were indicated by the use of a large R followed by 01, 02, etc, for the regimental

commander and his staff in declining order of importance.

As always there were many variations on this theme. On occasion four-digit codes were encountered, and some tanks carried two-digit or single-digit numbers. The four-digit numbers were usually applied to the reconnaissance units of large formations. Battalion headquarters tanks were often marked by the use of Roman numerals in place of the first number.

These numbers were applied in a wide range of sizes and styles. Colours used ranged from a simple black to white to yellow or red outlined in white. As well as being painted on to

Above *A battalion headquarters PzKpfw II Ausf B belonging to the 4th Panzer Division. The 106 denotes it is from the first battalion. Painted on the side of the turret is the name Hanz Lott, doubtless the name of a comrade killed in action who belonged to the same unit.* **Below** *A Hummel displaying its tactical sign on the front plate to the left of the spare wheels.*

Camouflage and markings

the turret sides, they were sometimes repeated on the turret rear or sides of the hull. The same system was often used by self-propelled artillery and assault guns when they formed part of a panzer formation.

Another tactical marking used on tanks was the tactical symbol or Taktische Zeichen. This was a small symbol painted on to the tank front and rear for the guidance of traffic police and others arms as the exact function and tactical seniority of the vehicle to which it was applied. These symbols were the same as those used on tactical maps and were usually very simple outlines in white or yellow (sometimes red was used) and were designed to be instantly recognisable.

As so often happens, a simple idea was soon made complex by the addition of flags, etc, to the basic symbol to denote the rank of the user, type of

Top row, left to right *1st Panzer Division 1940. Shown in the style used during the 1940 French campaign. The 2nd Panzer Division used two Xs side by side. 5th Panzer Division 1940. The 6th Panzer Division used the same symbol with two 'balls'. 4th Panzer Division 1941-45. The inverted Y was*

used also by the 1st (no bars), 2nd (one bar), 3rd (two bars) and 4th Panzer Divisions. The 7th to 10th Panzer Divisions used the same sequence with an upright Y. 3rd Panzer Division 1940. The circular symbol was used by the 3rd, 4th (plain upright Y) and Rommel's 7th Panzer Division (plain upright Y). **Second row, left to right** 15th Panzer Division 1941-43. This unique marking was used by the original Afrika Korps panzer unit, and later by the 15th Panzergrenadier Division. 18th Panzer Division 1941-45. Similar were the 16th (one bar) and 17th (two bars) Panzer Divisions. 20th Panzer Division 1943. As used in Central Russia, 1943. 21st Panzer Division 1941-45. **Third row, left to right** Gross-Deutschland Panzer Division. This symbol was exclusive to the elite Gross-Deutschland Division and its breakaway units. 1st, 2nd and 3rd SS Panzergrenadier Divisions. These symbols were used during 'Operation Citadel' at Kursk in 1943, and were deliberately changed to confuse the Russians. **Fourth row, left to right** 18th Panzergrenadier Division. Panzer-Lehr Division. 3rd Panzergrenadiar Division. 10th Panzergrenadier Division.

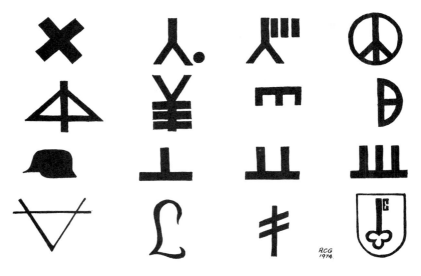

German tanks of World War 2

Above *This Hornisse shows both a regimental badge (unidentified) and a tactical symbol on its track fenders.* **Below** *A good example of the placing of markings on a Panzerjäger für 7.5 cm Pak 40 (Sf) Lorraine Schlepper. Often known as the Marder I this vehicle was the conversion of a French artillery tractor to carry a German gun. The unit marking may be that of the 320 Infantrie Division.*

Camouflage and markings

Top *4.7 cm Pak(t) auf PzKpfw 35R(f). These vehicles were German impro-
visations on a French Renault R35 tank chassis to carry a Czech anti-tank gun.
The unit marking is uncertain but may belong to the 21st Luftwaffe Field Division.* **Above** *This PzKpfw III Ausf G in Yugoslavia has the tactical code 415 on the
hull side. Just visible is the badge of the 2nd Panzer Division by the driver's visor.
The vehicle behind is a PzKpfw III Ausf H.*

armament carried, and so forth, so
that the basic simple idea became
complex and cumbersome in use. For
tanks the basic symbol was a rhom-
boid, but it was not often carried, or
was often painted out. However, the
system was widely used on other types
of vehicle used in panzer formations.

Above PzKpfw III Ausf E or G in Russia, 1941. The tank in the foreground has 14K painted on the spare turret pannier secured to the hull top. The K denotes that it is attached to the Panzer Korps commanded by Kleist. **Below** A Panzer III Ausf L of the 7th Panzer Division advancing into unoccupied France in 1942. The divisional symbol (Y) can be seen on the turret bin.

Camouflage and markings

Above *A captured PzKpfw III Ausf G displaying the divisional sign of the Afrika Korps, although strictly speaking this was not a division formation.* **Right** *The Afrika Korps symbol on a PzKpfw IV Ausf D.*

A PzKpfw IV Ausf H of the 1st SS Panzer Division. SS divisional markings were usually depicted inside a shield with a 'scalloped' top right hand corner — colours were usually white. Throughout the war the SS formations usually obtained a priority in the issue of men and equipment, and were often the first to receive modern vehicles.

German tanks of World War 2

Divisional signs

Every panzer division used some form of divisional sign which was painted on to the front and rear of their tanks. These signs ranged from the simple to the complex. The first panzer divisions used very simple signs made up from straight lines only, and they were very easy to apply, remember and recognise. Later signs were more complex. The simple signs were painted on to the vehicle with white or yellow paint, and later signs often used a variety of colours. After about 1944 it was not unusual to see tanks without divisional markings, for after that time tanks were used less and less in divisional formations and more and more in *ad hoc* battle groups (Kampfgruppe) formed for specific tasks.

Other markings

This sections covers a wide and varied range. German tanks often carried a variety of personal or unit good-luck symbols or signs. Some tanks were given names by their crews, or were named after wives or girl-friends, but this practice was officially frowned upon. Tanks rarely carried the vehicle number plates used by all the other transport vehicles in the Wehrmacht.

seven

Basic book list

As this short book has been intended as a primer only, many readers will want to delve further into the various facets of German tank history. There are many books on the subject available and the list below is by no means complete.

German Tanks of World War 2, by Dr F. M. von Senger und Etterlin. Arms and Armour Press. Translated into English, this book has long been the standard work on the subject. Well illustrated with many drawings.

Pictorial History of Tanks of the World 1915-45, by Peter Chamberlain and Chris Ellis. Arms and Armour Press. This book not only lists German tanks but tanks of all nations and thus puts the German designs into their correct context for comparisons. Each tank gets at least one illustration, many of them very rare.

Panzer Leader, by H. Guderian, now available in paperback. A near-official account of the rise and fall of the panzer forces written by the man who was in control of them during much of the war.

Panzer Division, by K. J. Macksey. Published in paperback as part of Purnell's History of the Second World War — Weapons Book No 2. An easily readable account of the panzer units' fortunes from start to finish. Well illustrated with many clear maps.

Wehrmacht Camouflage and Markings 1939-1945, by W. J. K. Davies. Almark Publications. A well-illustrated work with greater detail on the many variations used on German tanks and other vehicles.

German Army Handbook, by W. J. K. Davies. Ian Allan. A useful primer for setting the panzer organisations into the framework of the German army.

Design and Development of Fighting Vehicles, by R. M. Ogorkiewicz. Macdonald or Arms and Armour Press. Anyone wanting to learn more about the general design factors affecting tanks will do well to read this not-too-technical work.

Armoured Forces, by R. M. Orgorkiewicz. Arms and Armour Press. While not dealing exclusively with German tanks, this excellent book outlines the development of the tank as a fighting weapon from its birth to the present day. A must for any tank enthusiast.

Profile Publications. There are several Profiles dealing in detail with various German tanks, but they are all combined in Volume 5 of the Armoured Fighting Vehicles of the World series. The Profile series is well produced and written, with many good illustrations and colour spreads.

Tanks of other Nations — Germany 1917-1968. Available from the RAC Tank Museum, Bovington Camp, Dorset. A useful little booklet from the museum where many German tanks are preserved for inspection.

There are many modelling magazines on the market which sometimes give space to articles and features on German tanks. *Airfix Magazine* is an obvious example. Various specialist magazines are published by such societies as the Miniature Armoured Fighting Vehicle Association, who publish *Tankette.* This small magazine features modelling drawings and features on a wide variety of vehicles, and German AFVs are often included. Details can be obtained from G. Williams, 15 Berwick Avenue, Heaton Mersey, Stockport, Cheshire.